JERSEY GE
TRAIL

A SERIES OF EXPLANATORY NOTES ON SELECTED TRAVERSES

R.A.H. Nichols, Ph.D. and A. E. Hill, Ph.D.

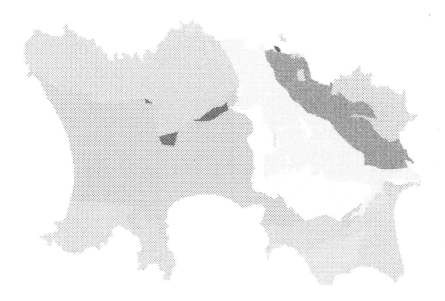

Traverses 1 to 6

To be used in conjunction with the IGS *Channel Island Sheet 2*, 1982 and the BGS *Classical Areas of British Geology: Jersey*, 1989.

Jersey Geology Trail

First published in Great Britain
in 2016 by Charonia Media

This is a reproduction (with minor alterations) of
Jersey Geology Trail, a book that was privately published
by its authors in 2004.

ISBN 978-0-9560655-3-7

CONTENTS

List of Figures

List of Maps and Diagrams

Bibliography

Bishop, A. C. & Bisson, G., *Classical Areas of British Geology: Jersey.* British Geological Survey, 1989.

Bishop, A. C. & Keen, D. H., *Geology of Jersey.* Geologists Association Guide No. 41, 1981.

D'Lemos, R. S., Strachan, R. A., and Topley, C. G., (Eds) *The Cadomian Orogeny.* Geological Society Special Publication, No. 51, 1990.

IGS, *Jersey: Channel Islands Sheet 2, Solid and Drift (1:25,000).* Institue of Geological Sciences, 1982.

Mourant, A. E., The Minerals of Jersey, *Annual Bulletin Société Jersiaise,* 1961.

Nichols, R. A. H., and Blampied, S., *Jersey's Geological Heritage: Sites of Special Interest.* Société Jersiaise, 2016.

Ordnance Survey, *Official Leisure Map, Jersey.* Ordnance Survey, 1988 (reprinted 1999).

Further reading and sources are obtainable from:
Société Jersiaise Book Shop, 7 Pier Road,
St Helier, Jersey JE2 4XW
www.societe-jersiaise.org

Acknowledgements

The authors gratefully acknowledge the discussions with members of the Geology Section of the Société Jersiaise and, in particular, the note taking by Deirdre Shute during our field visits. The photographs are the sole property of Dr A. E. Hill.

Key to
Geological sketch map of Jersey

Rozel Conglomerate Formation

Granites and granophyres

Diorites and gabbros

Bouley Rhyolite Formation

St. John's Rhyolite Formation

L'Homme Mort Conglomerate

St. Saviour's Andesite Formation

Jersey Shale Formation

Conglomerate in
    Jersey Shale formation

Traverse number and line

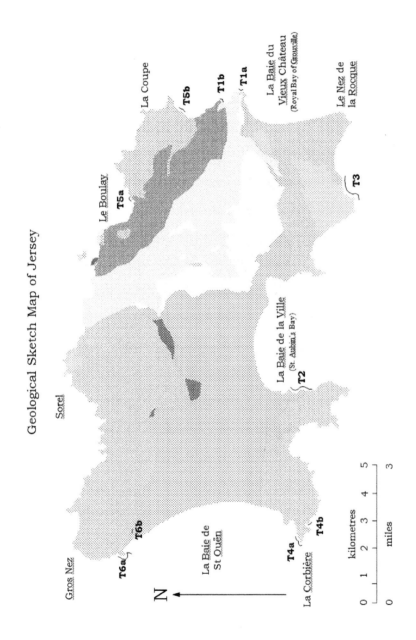

Geological Sketch Map of Jersey

THE REGIONAL CONTEXT OF JERSEY SOLID GEOLOGY

Jersey is situated in the Gulf of St Malo between Normandy and Brittany. It seems to have been largely unaffected by Palaeozoic deposition and by the Caledonian and Variscan orogenies. Thus the island consists mainly of Precambrian to Lower Palaeozoic shale, andesite, rhyolite, ignimbrite, and conglomerate formations which were folded by the Cadomian orogeny and then intruded by gabbro and granites of differing Lower Palaeozoic ages. Diorites were produced by alteration of the gabbro. Consequently, there is a record of sedimentary, igneous, tectonic, and some metamorphic activity from about 900 to 400 million years ago.

Geologically, Jersey forms part of the type area of the North Armorican Massif of France. The rocks range in age from late Precambrian to early Palaeozoic, approximately 750 to 425 Ma, and they were affected by tectonic and thermal events during the Cadomian orogeny (named after Cadomus, the Roman name for Caen in Normandy). These events were broadly coeval with those which affected the Avalonian rocks of southern Britain and the Monian rocks of NW Wales and SE Ireland (D'Lemos *et al*, 1990, p.3).

The oldest rocks exposed constitute the Jersey Shale Formation (Brioverian in age, *c.* 680 to 570 Ma) and consist of various ripple-laminated, cross bedded and graded, fine grained sandstones, greywackes and some mudstones. They were deposited in a late orogenic basin in several different, submarine deltaic fan environments subject to turbidity currents (Bishop and Bisson, 1989, p.9) and they have been regionally metamorphosed to low green schist facies.

These are overlain disconformably and conformably in parts by andesites, rhyolites and ignimbrites of the St Saviour's Andesite, St John's Rhyolite and Bouley Rhyolite Formations respectively, which form part of a second phase of Cadomian vulcanism from about 533 Ma ago.

These formations are thought to have developed in a late orogenic basin formed in the earlier orogenic belt which was a result of subduction of the Celtic Oceanic Crust southwards towards the Armorican Massif continental margin (Rabu *et al.* in D'Lemos *et al.*, 1990, p.89). Variation in strike directions indicate that deformation appears to have happened several times between 600 to 540 Ma to fold the shales and volcanic rocks about axes varying from N - S to E - W.

The shales and volcanic rocks were intruded during an extension phase by several plutonic complexes, respectively the SW, NW and SE granites varying in age from 550 to 425 Ma. The SE complex cannot be dated precisely but may be coeval with the similar NW complex. They represent mixed magma activity and vary from gabbro through diorite to granite. Geochemical studies suggest they were derived from mantle wedge melts with assimilation and fractional crystallisation at the base of the crust and during emplacement (Brown *et al.* in D'Lemos *et al.*, 1990, p.197). None of the complexes is foliated thus indicating post tectonic formation. Granite surveys imply that the granites, gabbroic in part, are continuous across the island (Bishop and Bisson, 1989, pp.97-102)

Subsequently, from 480 to 427 Ma, many dykes were intruded, mainly dolerites but with some mica lamprophyres and quartzo-feldspathic varieties. They were altered to low green schist facies metamorphic grade. These form the Jersey Main Dyke Swarm which cuts across the Jersey Shale and the post orogenic plutonic complexes along joints and some faults. This intrusive phase is also thought to be a result of post orogenic crustal extension. They strike about E - W directions, sub parallel to the proposed Cadomian active continental margin. Geochemically, they are similar to those of presently active continental margins, e.g. the Andes (Lees in D'Lemos *et al.*, 1990, p.273). They were later faulted by both sinistral and dextral tear faults.

The NE corner of Jersey comprises the Rozel Conglomerate Formation resting unconformably on the Bouley Rhyolite Formation. It consists of clasts of older rock types of the Jersey Shale and volcanic units and of granites which are not of Jersey granite composition. Some red silt and sandstone beds occur and the Formation seems to have been deposited under semi-arid conditions in an E - W striking inter montane basin as a molasse deposit of alluvial fan type during Cambro-Ordovician times (Bishop and Bisson, 1989, p.41).

The States of Jersey recently designated 22 of the island's most important geological sites as **Sites of Special Interest** (SSIs) so that they may be protected from development and preserved for public enjoyment and research purposes. The reader should be aware that some of the locations described in this guide are protected geological SSIs (see page 134). For more information see Nichols and Blampied (2016) or the States of Jersey website.

SAFETY NOTE

Some of the descents are down cliffs; it may be better to choose a route to the land side of the fishermen's paths which often pass very close to the cliff edges. Much of the walking at sea level is over loose pebbles and boulders, and outcrops which may be covered in slippery sea weeds. Care should be taken and the appropriate footwear used for these conditions. Do watch the tides, and always consult the local tide tables for Jersey which has a very large tidal range.

Explanatory Notes No. 1a &1b

Le Petit Portelet, Gouray and La Crête Point, Anne Port

Ordnance Survey Grid References
71555048 and 71405120

CARE SHOULD BE TAKEN TO CHECK TIDE TIMES AT BOTH SITES.

The duration of each visit should be about one hour per site.

The geological sketch map T1 may be used to locate the various points of interest by using the positions of the Figure numbers,
e.g. Fig. 1 refers to Site 1 on the map.

Key to Geological Sketch Map T1

Anne Port Rhyolite

Ignimbrites

St. John's Rhyolite Formation

Bouley Rhyolite Formation

Jeffrey's Leap Ignimbrite

St. Saviour's Andesite Formation

La Rocque granite

Jersey Shale Formation

Faults
Inferred faults
Inferred geological boundaries
Mica Lamprophyre dykes
Coast line
Cliff line

Rock stack

Roads
Footpaths
CP   Car Park
PH   Public House

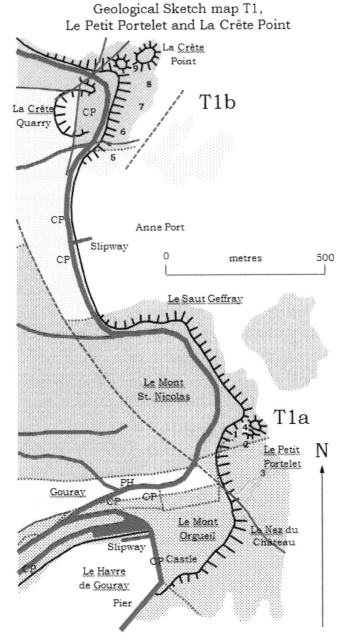

Geological Sketch map T1,
Le Petit Portelet and La Crête Point

Geological Sites of Special Interest

Please note that this trail includes outcrops(s) that are protected as Sites of Special Interest (SSIs) under Jersey's Building and Planning Law (2002). Please see page 134 in this book for more details.

T1a - Le Petit Portelet

This is a small bay situated immediately north of Le Mont Orgueil and Gorey castle on the east coast of Jersey. It is approached by a gentle path to the beach from Castle Green, and should be visited from mid to low tide. There is a pebble beach consisting of a great variety of Jersey rock types but the main interest lies in three separate outcrops which are situated along the beach from in front of the viewer southwards towards the castle.

Figure 1. Ignimbrite structure at Petit Portelet (x 0.10)

The first outcrop consists of Jeffrey's Leap Ignimbrite, part of the St John's Rhyolite Formation. It is superficially a red brown rock, well jointed, with little evidence of layering. Surface examination reveals the streaked (eutaxitic) texture of a finely crystalline, felsitic ignimbrite with scattered angular fragments of pale ash and dark xenoliths of country rock, dipping east to north east (**Figure 1**).

The second outcrop southwards is interesting because it shows the ignimbrite contact with the La Rocque granite forming Le Mont Orgueil. This seems to be a well defined contact in part exhibiting various textures, possibly due to temperature differences, and a faulted contact elsewhere with some brecciation. The granite is red brown, medium crystalline with scattered white feldspars (**Figure 2**).

Figure 2. Ignimbrite/granite contact (x 0.13)

22

Figure 3. Bifurcated mica lamprophyre dyke (x 0.13)

The third and final outcrop is part of the main beach rock mass extending to below the castle. Here, at the northern end, the granite has been intruded by a dark brown, NE-SW striking, mica lamprophyre (minette) dyke which bifurcates into one about 1 m wide and another of about 30 to 40 cm width. They have been eroded to form gullies as they are softer than the granite; the contact walls are clearly exposed (**Figure 3**). The micas vary in size from 5 to 10 mm and are brown-bronze in colour. Within the dykes, which have been faulted in several places, by N - S striking, possibly sinistral tear faults, there are thin slivers of granite.

Another feature of interest is a narrow vein of baryte, white to pink in colour, 1 to 5 cm wide and striking SE - NW. This occurs in a crevice and gully in the big headland at the northern end of the beach (**Figure 4**).

Figure 4. Baryte vein (x 0.14)

Superficial deposits representing glacial head clays and rock fragments form the cliffs at the back of the beach, and recent pebbles showing a great variety of colours, textures and compositions lie below on the beach and around the outcrops.

T1b - La Crête

This is a point or headland situated some 2 km to the north of Gouray and forms the north side of Anne Port bay on the east coast of Jersey. The safest approach is from Anne Port slipway from mid to low tide, walking north east across the sand to the first, or southern most cliff outcrops which consist of St John's Rhyolite Formation ignimbrites and Bouley Rhyolite Formation flows.

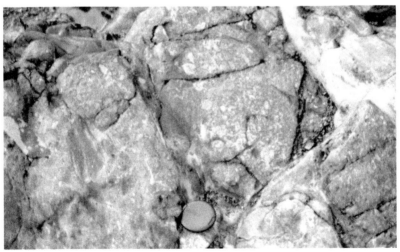

Figure 5. Ignimbrite texture in Anne Port Rhyolite (x 0.2)

In a wide gully immediately to the north of these cliffs there are excellent outcrops of the Anne Port Rhyolite which consists of five rhyolite flows overlying the Anne Port Ignimbrite. The first flow, in the gully, exhibits large scattered angular fragments in a finely crystalline groundmass (**Figure 5**). The fragments are of various coloured country rocks and minerals. This is overlain higher in the adjacent cliffs by red brown, obviously flow banded rhyolite, the upper part of which is autobrecciated (rubble-like due to weathering).

Figure 6. Gas bubble? (x 0.15)

In the next small embayment there are several layers of variable thickness, and formed of volcanic breccias dipping to the north east. They contain fragments varying from small, lapilli size pyroclasts to larger, angular agglomerate size ones. A thin andesite breccia (Anne Port Andesite) 30 to 50 cm thick, crops out at the base of the next low cliff to the north and contains an interesting pillow-like structure. It is not a submarine feature and is interpreted as a gas bubble (**Figure 6**) structure near the surface of the layer. The upper parts of the flows are brighter

Figure 7. La Chaussé de la Crête. The outcrop stands some 13 metres above mean sea level.

red and rubble-like and again represent surfaces weathered subaerially and penecontemporaneously to produce autobreccias.

Continuing north, one is confronted by an outcrop showing a very uniform structure of angular columns inclined away from the viewer. This is our own Giants' Causeway, (La Chaussée de la Crête) (**Figure 7**), a columnar jointed rhyolite flow with the columns varying irregularly from pentagonal to hexagonal in section, and seeming to be stacked at an angle leaning northwards. If formed vertically through the flow about cooling centres, they indicate a northerly dip to the flow varying between 30° and 60°. How much is tectonic and how much is original flank dip is unknown as the vent(s) have not been found.

Figure 8. Flow banded rhyolite columns (x 0.15).

In addition, this flow exhibits excellent white and red-brown flow folds and bands which cut across the columns (**Figure 8**), plus areas of small, 1 to 3 mm, white crenulated spherulites (**Figure 9**) and later quartz veins. In many cases the flow bands have been weathered to cause separation into discrete, thin layers across the columns. There is also a line of sigmoidal tension gashes lying en échelon to each other to be found in a gully at the top of the beach at the foot of the columns.

Finally, it is worthwhile examining the boulders along the seaward edge of the rocks and in the gullies as they provide good examples of local rocks, not always from the surrounding outcrops!

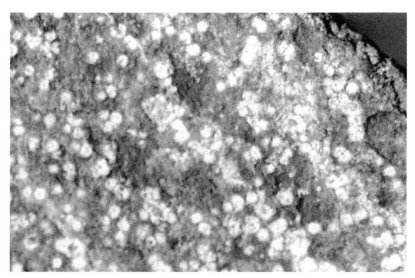

Figure 9. Crenulated spherulites (x 1.35).

In summary, the geological history of the area is as follows. Bursts of volcanic activity produced the various ignimbrites, volcanic breccias and banded flows. These were later intruded by granite which in turn was intruded by the lamprophyre dyke. Subsequent uplift, erosion and glaciation produced the glacial head and present day coastline.

Explanatory Notes No.2

St Aubin
to
Belcroute

Ordnance Survey Grid References
60734857 to 60754761

**CARE SHOULD BE TAKEN TO CHECK TIDE
TIMES AT BOTH SITES.**

The duration of the visit should be
about two hours.

The geological sketch map T2 may be used to locate
the various points of interest by using the positions
of the Figure numbers,
e.g. Site 10 on the map refers to Figure 10.

Key to Geological Sketch Map T2

SW granite (Corbière type)

Jersey Shale Formation

Faults
Inferred faults
Inferred geological boundaries
Mica Lamprophyre dykes
Rhyolite Dyke
Dolerite Dyke
Coast line
Cliff line

Rock stack

Roads
Footpaths
CP   Car Park
PH   Public House

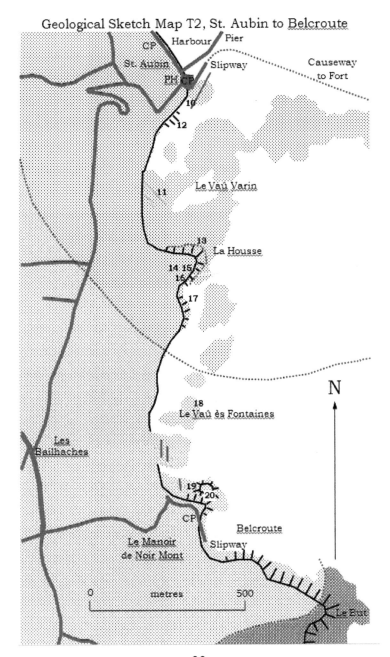

Geological Sketch Map T2, St. Aubin to Belcroute

Geological Sites of Special Interest

Please note that this trail includes outcrops(s) that are protected as Sites of Special Interest (SSIs) under Jersey's Building and Planning Law (2002). Please see page 134 in this book for more details.

T2 - St Aubin to Belcroute

St Aubin is situated at the western end of St Aubin's Bay on the south coast of Jersey, west of St Helier.

Figure 10. Laminated shales (x 0.13).

This well exposed traverse extends along the beach and cliffs from the slip-way (to St Aubin's Fort), southwards for about 1.2 km to Belcroute. The excellent outcrops consist of rocks forming part of the Jersey Shale Formation and the SW Granite (Corbière type). After uplift and folding the shales and granites were intruded by a variety of dykes, faulted, uplifted and eroded to produce the present coastline. Superficial deposits of glacial head and beach pebbles of different ages overlie an interesting erosion surface with angular unconformity.

Figure 11. Mica lamprophyre dyke (x 0.10)

The first site occurs on the south side of the slip-way. The outcrop comprises well jointed dark grey shales. They are laminated mudstones with ripple laminae of fine grained, light grey sandstone layers, about 1 to 2 cm thick. The various textures and structures such as cross bedding and grading are easily visible, being identified by light and dark laminae (**Figure 10**). These sets also illustrate the dip in this area which varies from 70° to 80° in a general SE direction. The laminated nature of the shales suggests that they were deposited as muds with turbidity silts and sands on the outer edges of a submarine deltaic fan (Bishop and Bisson, 1989, p.11).

Figure 12. Joint flexure in shale beds (x 0.10)

37

Here, they are intruded by a dolerite dyke about 1 m wide, striking E - W and faulted in part by apparently sinistral tear faults. Brecciation and chemical weathering to hydroxides have occurred along the faults and small gully erosion seems fault controlled.

The traverse continues southwards where isolated masses of similar shale crop out from Lé Vaû Varin and SE striking dykes can be seen; two narrow mica lamprophyre and a dolerite. The former (**Figure 11**) have been faulted by dextral N - S tear faults near the back of the beach where one or two examples of joint flexure (**Figure 12**) can also be seen.

Along the cliffs of glacial head there are scattered exposures of pebble beds representing a raised beach of variable extent and thickness.

Steps up the first headland lead to a clearly visible bed rock erosion surface overlain by glacial head similar at first sight to a standard soil profile, except for the yellow colour and the clearly defined lithology change showing incipient stratification (**Figure 13**).

In the shales the strike and dip directions remain the same and the steeply dipping bedding planes form the cliff faces.

Figure 13. Glacial head (x 0.05)

Figure 14. A rhyolite dyke (x 0.005)

A rare, flow banded, rhyolite with a number of interesting structures forms a dyke with E - W strike at the south end of the bridge and in beach outcrops (**Figures 14, 15 & 16**).

Figure 15. A rhyolite structure (x 0.10)

Figure 16. A rhyolite structure (x 0.10)

Walking carefully across the present erosion surface, with the appearance of a raised wave cut platform, leads down to a small bay with a sloping pebble beach, backed by high cliffs of yellow glacial head deposits, capped by trees.

These deposits overlie the shale bedrock, here eroded to form ridges and gullies filled with angular shale debris, overlain by the glacial head. Towards La Housse, the southern point of this bay, there is an excellent example of a pebble bed about 1 m thick and some 20 m wide, near the base of the head (**Figure 17**). It consists of oval, well rounded granite

Figure 17. Raised pebble bed (x 0.10)

pebbles and seems to thin out and then disappear north and south, its actual extent being obscured by beach pebbles and land-slips of head. This is regarded as part of a raised beach, locally called the 8 m beach. This site provides a good opportunity to review the geological history of the traverse thus far (**Diagram 1**).

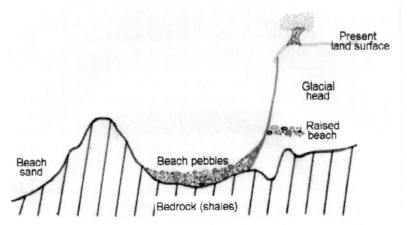

Diagram 1. An east - west sketch section suggesting geological history (not to scale)

South of this bay and point, another bay, Lé Vaû ès Fontaines, extends towards Belcroute. Here, there is a noticeable change in the beach pebbles and outcrop colours from grey to pink and red-brown. Careful study of the last grey outcrop, reveals the jointed, steeply dipping laminated mudstones, their textures still preserved, though some conversion to hornfels may have occurred. The next outcrops south, in the beach and at the back, are of the SW granite with medium crystalline white quartz and pink feldspars, some showing zoning. The contact is below the pebbles (**Figure 18**).

Figure 18. A hidden geological boundary between two rock outcrops.

At the back of the beach, above the sea wall, the 8 m raised beach is still evident, varying from 60 cm to 1 m thick and consisting of oval, rounded granite pebbles, mixed with angular fragments in some places. Here, the pebbles, are situated above a granite erosion surface, though not above small crags which may therefore represent former reefs. The granite outcrops from here to the last headland before Belcroute, have been intruded by three dolerite dykes varying from 1 to 2 m wide and eroded into pebble filled gullies seawards, thus revealing the steep sided contacts. Again, weathering has caused changes in mineral composition and colour.

The first dyke has been offset, west side northwards indicating dextral tear faulting. The second forms a small exposure some 2 m wide, but the third is wider, longer and is also tear faulted west side northwards (**Figure 19**); it occurs just before the headland on the south side of this bay. All three strike approximately ESE while the tear faults strike approximately NNE and dip towards the west.

Figure 19. A tear fault truncating an eroded dyke gully (x 0.04)

The final outcrop, the point itself, is of coarsely crystalline red granite with some narrow quartz veins and isolated patches of large milky quartz and pink feldspar crystals exhibiting pegmatite texture (**Figure 20**).

Figure 20. Orthoclase quartz pegmatite (x 1.0)

In summary, the geological history of the area is as follows. The Jersey Shale Formation (Brioverian) was deposited during the Precambrian, about 700 Ma ago, in a submarine deltaic environment. It was folded, uplifted and eroded during the Cadomian Orogeny, in late Precambrian and early Palaeozoic times about 600 Ma ago due to NW - SE compression and then intruded by the SW granite and flow banded rhyolite about 560 Ma ago. These two, granite and shale, were then intruded by dolerite and mica lamprophyre dykes along mainly E - W fractures as part of the Jersey Main Dyke Swarm. Faulting then caused the dykes to be offset by dextral and sinistral faults striking approximately NNE. Subsequent uplift and Pleistocene climatic and eustatic changes produced the raised beach deposits and glacial head, culminating in the present day landscape.

Explanatory Notes No.3

Le Hocq
to
Le Nez

Ordnance Survey Grid References
68494663 to 67784623

CARE SHOULD BE TAKEN TO
CHECK TIDE TIMES.

The duration of this visit should be about
three hours.

The geological sketch map T3 may be used to locate
the various points of interest by using the positions
of the Figure numbers,
e.g. Site 21 on the map refers to Figure 21.

Key to Geological Sketch Map T3

	La Rocque granite
	Diorites and Gabbros
	Faults
	Inferred faults
	Inferred geological boundaries
	Dolerite Dyke
	Feldspar Porphyry dyke
	Coast line
	Cliff line
	Rock stack
	Roads
	Footpaths
CP	Car Park
PH	Public House
SH	Summer House
T	Jersey Round Tower

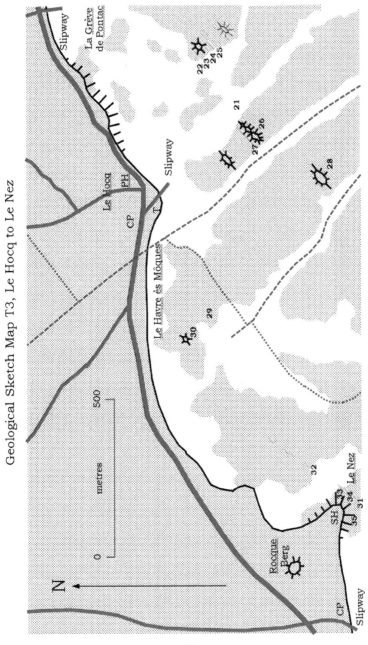

Geological Sketch Map T3, Le Hocq to Le Nez

Geological Sites of Special Interest

Please note that this trail includes outcrops(s) that are protected as Sites of Special Interest (SSIs) under Jersey's Building and Planning Law (2002). Please see page 134 in this book for more details.

T3 - Le Hocq to Le Nez

The area is situated on the south coast of Jersey and extends along the wave cut platform from the slip at Le Hocq westwards to Le Nez Point and to La Motte (Green Island) car park.

The different rock types are well exposed and form part of the SE granite and the SE Igneous Complex, a mixture of various gabbros, diorites and granites, intruded by many dolerite dykes which form the Jersey Main Dyke Swarm.

Figure 21. Granite stack on Le Hocq foreshore

The first outcrops to be examined occur some 200 m ESE of the slip-way and form reefs and low stacks of red-pink, medium to coarse crystalline La Rocque granite which also contains porphyritic microgranite (**Figure 21**). They are well jointed, though not at 90°, with major planes striking ENE, and steeply inclined. The most noticeable feature is the swarm of more than fifteen medium grey, dolerite dykes generally varying in thickness from 0.6 to 3 m which strike NE and dip at about 80° to the SE. They are parallel to the major joints and have been intruded along them (**Figure 22**). Some of them have

Figure 22. Granite crag with dolerite dykes.

been displaced by SSE striking sinistral tear faults. Other thin dykes are generally 2 to 3 cm thick but may thin and cut out, while some show bifurcation with central granite lenses and others remain single.

Green epidote mineralisation is present on some contact surfaces (**Figure 23**). Joints occur within the dykes with some of them crossing the dyke into the granite either side (**Figure 24**).

Figure 23. Epidote mineralisation (x 0.23)

Figure 24. Joints crossing dolerite dyke and host granite.

To the south of the stack is a large dolerite dyke which shows both bifurcation and curved strike without apparent faulting (**Figure 25**).

Figure 25. large bifurcated dolerite dyke with curved branch (x 0.05).

Aplite veins penetrate the granite. These vary in width from 1 to 10 cm, are jointed and weathered to give differences in colour and outline.

Westwards, the triple stack is of granite, again intruded by dykes. On the south side a sinistral tear fault shows a decrease in throw from 1 to 5 m along 3.5 m of its SE strike. Aplite veins also occur, 1 to 10 cm wide, one with a thin brown weathered surface (**Figure 26**).

Figure 26. Weathered aplite vein (x 0.33).

On the next stack SW, which has been quarried, several dykes occur. On the north side, the main one has a central granite lens, a basic margin in part, and a decomposed centre (**Figure 27**).

Two dykes occur on the south side of the stack, one of which is faulted and disappears. Rare pegmatite patches (*c.* 60 by 60 cm) of large white quartz and red orthoclase feldspar intergrowths can also be seen (**Figure 28**).

Figure 27. Structure in dyke margin (x 0.17).

Turning NW towards the beach wall and the houses, the outcrops are lower and grey in colour. They represent basic and intermediate gabbros and diorites, which vary in colour from dark grey and black, to black and white speckled respectively, with yellow veins and patches of feldspar.

Some diorites appear brecciated or to contain xenoliths of gabbro where coarsely crystalline diorite veins have intruded them. Others, called appinites, have large black, euhedral, columnar hornblende crystals, 3 cm long, showing hexagonal sections and occur with anhedral, white and yellow, plagioclase feldspars (**Figure 29**).

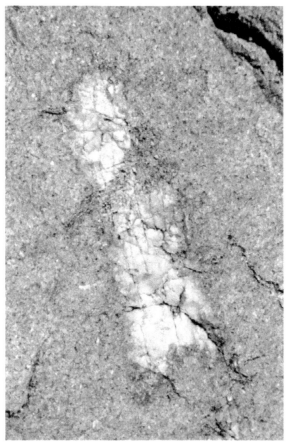

Figure 28. Quartz pegmatite (x 0.2).

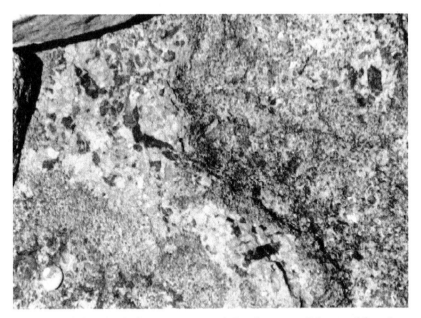

Figure 29. Appinite texture with elongated horneblende crystals (x 0.33).

In the outcrops nearer the beach wall below the houses, the outcrops vary from light grey, well jointed diorite (leucodiorite) with granite dyke intrusions, *c.* 400 m west of Le Hocq Tower, to dark grey-black, mixed diorite and gabbro (or melanodiorite) further west. In successive outcrops towards Rocque Berg and Le Nez Point, excellent examples of various different textures appear to illustrate mixing and segregation layering. Yellow-brown and white feldspar and black hornblende and augite layers grade into blacker, gabbroic layers. Veins, dykes and areas of feldspar occur producing a brecciated or xenolithic appearance.

Red and white crystalline veins and areas produce granitic looking textures. Yellow-brown and red colours are due to the hydrolysis and oxidation of iron. These textures may signify mobilisation, mixing and segregation during later intrusion of granitic material (**Figure 30**).

Figure 30. Mixed rock types (x 0.07).

A detour southwards to the last obvious red granite stack, reveals the irregular, jagged intrusive contact of the diorite/gabbro and the granite in the last low outcrop before the stack.

Proceeding to the east side of Le Nez Point, an inlet reveals a basic dyke striking ENE displacing layered, segregated diorite in light and dark bands with apparent dextral tear. Nearby, sharply defined veins and narrow dykes indicate greater temperature differences between the country rock and the intrusions. Two faults strike through here to the WNW. Immediately south of this inlet (below the summerhouse on the cliff top), the outcrop is intruded by a brown, feldspar porphyry (or porphyritic microgranite) dyke which strikes NE and is up to 3 m wide (**Figure 31**). It can be traced for 500 to 600 m eastwards towards Le Hocq Tower, offset slightly in places by tear faults below pebble filled gullies.

Figure 31. Feldspar porphyry dyke (x 0.015).

In the first low outcrop eastwards by some 10 metres there is an interesting structure. On the south side of the brown dyke, thin light and dark layers in the diorite exhibit a rectangular pattern and follow 90° joints, therefore seeming to be controlled by them, i.e. post dating them (**Figure 32**).

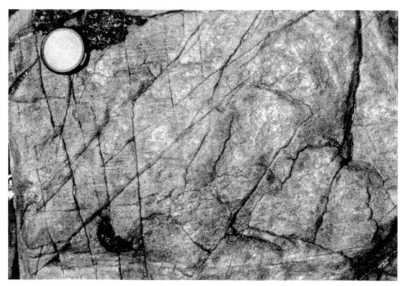

Figure 32. Rectangular jointing and layering in diorite (x 0.20).

Back on Le Nez Point outcrop, a small cliff of diorite exhibits a web of light coloured quartz and feldspar veins and elsewhere the diorite has dark grey-black gabbroic areas, lenses and layers together with pink and white quartzo-feldspathic (granitic) lenses and layers. This could represent an area of stoping or of quartzo-feldspathic intrusion (**Figure 33**).

Figure 33. Quartzo-feldspathic intrusion (x 0.1).

Down at sand level, blocks broken from the crags above, exhibit areas of black and white speckled diorite and large yellow-white feldspar laths, 2 to 3 cm long, with large black hornblende crystals, some 10 cm long, some showing hexagonal sections (**Figure 34**). Alternating feldspar and hornblende layers may represent the segregated edges of a molten pocket.

Round the corner westwards again there are light and dark diorites with pink and white quartzo-feldspathic areas and veins. Layers of light and dark diorite 2 to 5 cm thick with thin central quartzo-feldspathic veins dip seawards, while a complex dyke, possibly showing multiple intrusion with granitic lenses and veins, some fault-truncated, strikes into the sea wall behind (**Figure 35**).

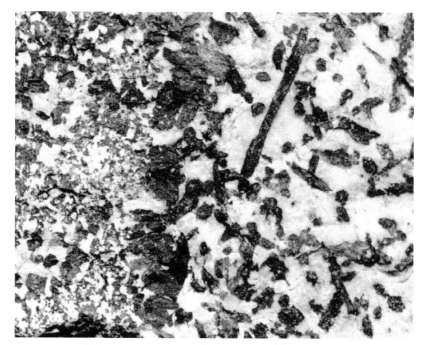

Figure 34. Hornblende crystals (x 1.7).

South of the car park for La Motte (Green Island) the first beach outcrop is of mixed diorite and gabbro in light and dark layers with narrow quartzo-feldspathic veins.

Figure 35. Complex dyke (x 0.03).

In summary, no reliable absolute ages have been determined for this area but the sequence seems to start with the formation of the gabbroic part of the SE Igneous Complex. This was later intruded by the La Rocque granite which in places seems to have caused the formation of the diorites by assimilation and then their segregation, when cool enough. However, in other places the granite seems to have intruded a cooler gabbro judging by the sharper defined contacts. The complex was then intruded by the dolerite dykes (the Jersey Main Dyke Swarm) along well defined joints in the granite; these were later jointed and faulted by sinistral and dextral tear faults. Evidence of low grade regional metamorphism exists in the presence of prehnite-pumpellyite minerals. These are green, hydrated calcium, and iron magnesium, aluminium silicates, alteration products from calcic plagioclase, and augite and hornblende respectively.

Explanatory Notes
No.4a and 4b

Le Grouët slipway
to
La Corbière causeway,
and
La Rosière.

Ordnance Survey Grid References
55514836 to 55084809, and 55734793

CARE SHOULD BE TAKEN TO CHECK TIDE TIMES
AT BOTH SITES.

The duration of the first visit can be some two
hours.

A period of half an hour may be allowed for the
second site not taking into account the time taken
to get from one site to the other.

The geological sketch map T4 may be used to locate
the various points of interest by using the positions
of the Figure numbers,
e.g. Site 36 on the map refers to Figure 36.

Key to Geological Sketch Map T4

	Corbière granite
M	Magnetite
Q	Quartz pegmatites
	Faults
	Inferred Fault
	Dolerite dykes
	Coast line
	Cliff line
	Rock stack
	Roads
	Causeway
	Footpaths
CP	Car Park
PH	Public House
LH	Lighthouse
RT	Radio Tower

Geological Sketch Map T4,
Le Grouët slipway to La Corbière causeway, and La Rosière.

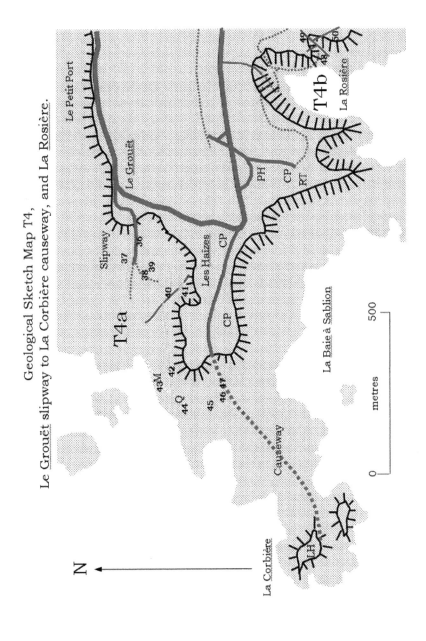

T4a - Le Grouët slipway to La Corbière causeway

The area is situated in the south west part of the island and extends westwards along the wave cut platform on the north side of La Corbière Point from Le Grouët slipway to the lighthouse causeway. The

Figure 36. Aplite vein (x 0.06).

outcrops consist of red, well jointed coarse crystalline granite intruded by dolerite and aplite dykes and rare mineral veins.

Alongside Le Grouët slipway and on the platform, the coarse granite is porphyritic in parts with large white quartz and red orthoclase feldspar crystals. Narrow, darker red, finely crystalline, vertical aplite dykes, striking E and ENE, are some 15 cm wide (**Figure 36**) and occasionally bifurcated.

Figure 37. Dolerite dyke (x 0.01).

The slipway continues as a sand gully and track-way through the granite. On the north side, a mass of dark grey dolerite with epidote on joint faces crops out as part of a dyke (**Figure 37**). The surrounding granite is medium grey in colour and may represent an intermediate or more dioritic rock due to contamination by basic doleritic magma. Wider aplite dykes can be seen south of the track.

Continuing along this route, pods and enclaves of aplite also occur as intrusions in the granite on the north side. Nearby, too, are more isolated masses of dolerite which appear to form part of a main dyke, some 3 m wide, striking E and ENE with differential erosion having caused the discontinuous outcrops.

Figure 38. Quartzo-feldspathic pegmatite lens (x 0.09)

A few metres further on, the track crosses a red and white quartzo-feldspathic pegmatite lens some 5 m long, 30 cm wide and sinuous in form (**Figure 38**). In addition there are lenses, small veins and irregular masses of dark grey dolerite within the granite (**Figure 39**), possibly indicating irregular intrusion at this level.

Figure 39. Dolerite in granite (x 0.03).

Further WSW, the traverse is cut by a 2 to 3 m wide gully striking SE, with narrow dolerite outcrops on the western edge showing variation to quartz rich margins in parts. The gully, partly filled with pebbles and sand, seems to represent an area where an ENE striking dolerite dyke has changed strike and/or been dextrally tear faulted some 30 m east side south-eastwards. Grey to pink granite forms the floor (**Figure 40**) of the gully northwards where it looks like another track-way.

Figure 40. Grey and pinkgranites (x 0.27).

Further west on the south side of the dyke, darker grey granite occurs with red, iron-rich feldspars and discrete quartz crystals in places and some dolerite exposures which appear to bifurcate or are partly sheared. Towards the house on the

Figure 41. Quartz feldspar granite (x 0.06).

cliff there are excellent areas of pegmatite (**Figure 41**), some 2 m long, with intergrowths of large milky quartz and red orthoclase feldspar crystals.

Turning north leads to a 3 to 4 m wide, differentially eroded dolerite dyke, with deep gullies along its edges, striking ENE (**Figure 42**). It is well jointed with some conjugate sets and is brown and grey mottled in parts due to variable weathering of the ferromagnesian minerals.

Figure 42. Differentially eroded dolerite dyke (x 0.01).

To the north again, the granite is intruded by a narrow quartzo-feldspathic dyke. Here the granite varies from red to grey in colour and with careful searching, discontinuous veins and small patches of dark grey magnetite can be found (**Figure 43**) and tested by its effect on the compass needle.

Figure 43. Magnetite locality (x0.01).

Figure 44. Large quartz orthoclase pegmatite (x 0.02).

Crossing back over the dyke and taking the highest route over the granite masses towards the lighthouse leads to a point overlooking a very large quartz pegmatite (**Figure 44**). Moving eastwards there are darker grey areas in the granite, composed of more hornblende. Narrow, red, finely crystalline aplite dykes strike N - S across this section and others, brown red in colour, can be found in the low cliff outcrops where the causeway flattens out towards the lighthouse.

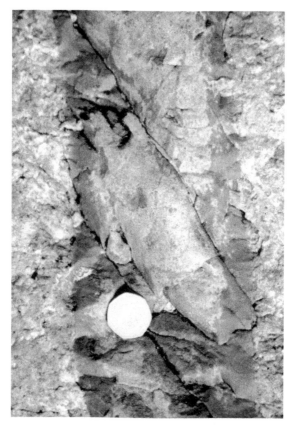

Figure 45. Aplite vein with selvages (x 0.35).

One such dyke shows clear alteration along its edges (**Figure 45**). Close to the causeway, a narrow, vertical, aplite dyke appears to act as a feeder dyke (**Figure 46**), but cuts across a near horizontal one which can be followed around the outcrops. (**Figure 47**) Here too, the granite shows colour changes in zones from pink to light grey, possibly representing segregation layers, seemingly due to variations in the hornblende content. Careful searching along adjacent fault planes may reveal small, weathered, iron pyrite crystals.

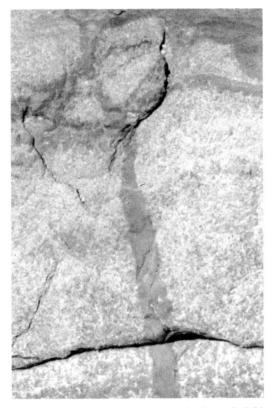

Figure 46. Aplite feeder vein? (x 0.08).

The coarse, crystalline granite is one of three granites forming the SW granite and is the youngest, at some 483 Ma (Rb-Sr whole rock isochron age). It was later intruded by several dolerite and aplite dykes along an ENE strike. The well jointed granite implies it was cool and solid before intrusion of the dykes forming the western part of the Jersey Main Dyke Swarm which are much younger than it (Lees in D'Lemos *et al.*, 1990, p.274).

Figure 47. Vein junction seen in Figure 46 (x 0.26)

T4b - La Rosière

The traverse can be extended into La Rosière, the next bay to the east of La Corbière, by following the cliff path from the car park next to the Radio Tower. The path descends via steps into the bay and passes round the headland to the south via a small quarry. A large aplite dyke will be crossed once leaving the bay and once again round the headland (**Figure 48**). Both the path and the dyke are truncated by a large gully that leads into a cave to the east That the gully was produced by erosion along a tear fault is indicated by the absence of the dyke on the opposite side (**Figure 49**).

Figure 48. large aplite dyke (x 0.02).

Figure 49. large gully in granite (x 0.006).

An intriguing plano-convex lens-like structure may be seen in the opposite cliff face near the top of the splash zone just above a horizontal rock shelf. This intrusion of aplite into the granite does not have a visible feeder (**Figure 50**). On the reverse, hidden, side of the stack the lens is thicker and contains larger amounts of white quartz. Again there is a fault to be seen here as the lens does not continue across the hidden gulley. It can be shown that the main gully has an eight metre sinistral fault associated with it by tracing the earlier fault and the aplite dyke across the large gully shown in **Figure 49**.

Figure 50. Aplite lens seen right centre in Figure 49 (x 0.01).

Explanatory Notes
No.5a and 5b

La Tête des Hougues,
and
Belval
to
Le Creux au Diable.

Ordnance Survey Grid References
67105443, and 71085283 to 71155290

**CARE SHOULD BE TAKEN TO CHECK TIDE
TIMES AT BOTH SITES.**

The duration of the first visit can be
some two hours.

A period of an hour may be allowed for the second
site not taking into account the time taken to get
from one site to the other.

The geological sketch maps T5a and T5b may be
used to locate the various points of interest by
using the position of the Figure numbers,
e.g. Site 51 on the map refers to Figure 51.

Key to Geological Sketch Map T5

Bouley Rhyolite Formation

Rozel Conglomerate Formation

Faults
Inferred faults
Inferred geological boundaries

Coast line
Cliff line

Rock stack

Roads
Footpaths
FB   Footbridge
CP   Car Park
PH   Public House
C   Café

Geological Sketch Map T5a, La Tête des Hougues

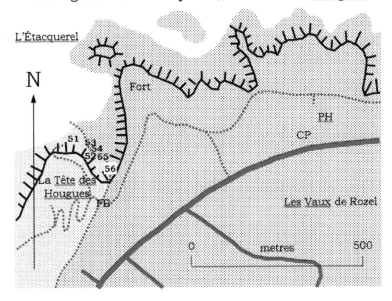

Geological Sketch Map T5b, Belval to Le Creux au Diable.

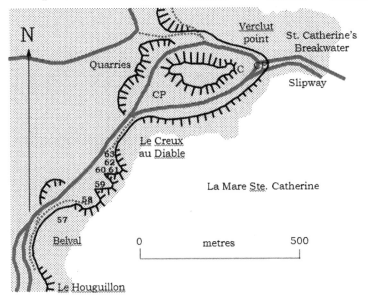

Geological Sites of Special Interest

Please note that this trail includes outcrops(s) that are protected as Sites of Special Interest (SSIs) under Jersey's Building and Planning Law (2002). Please see page 134 in this book for more details.

T5a - La Tête des Hougues

This traverse is situated in the Parish of Trinity on the north coast of Jersey, midway along the shoreline of Bouley Bay, in a small cove SW of L'Étaquerel point. La Tête des Hougues is a prominent hill of rock above the cliffs. It is approached along a farm track from La Route des Côtes du Nord (C93), OS Ref 679543, then by the cliff path round the east side of La Tête des Hougues and then, just after a plank bridge, right across the headland and down to the beach rocks via a fisherman's path.

Figure 51. Middle Bouley Ignimbrite (x 0.15)

Figure 52. Spherulitic quartz nodule (x 0.38)

The first outcrops are of the Middle Bouley Ignimbrite. They are dark brown-red to grey in colour, variously flow banded, nodular, ropey, and brecciated in appearance (**Figure 51**). The flow banding represents eutaxitic texture, the flattened, drawn out shards and pumice producing banding. The nodules are large spherulites, 1 to 4 cm in diameter, and occur as discontinuous layers >1 m thick and 2 to 3 m long or wide. Nearby, the spherulites consist of alternating, concentric bands of radiating fibrous feldspar, illustrating successive stages of growth (Bishop and Bisson, 1989, p.32). The ropey structures represent small flow folds.

Eastwards, as the coast turns into the small cove, these exposures become a sloping planar surface which dips irregularly to the NE and NNE at 25 to 30°. On this sloping surface, large spherulitic quartz nodules (**Figure 52**) can also be seen, along with narrow quartz veins striking NNE, making this outcrop part of the ignimbrite. It is here that one can see the basal beds of the Rozel Conglomerate resting on the eroded upper surface of the Middle and Lower Bouley Ignimbrite as reported by Renouf (Bishop and Bisson, 1988, p.42).

Figure 53. Site of the Rozel Conglomerate base.

It is an interesting exposure because in places it is not easy to distinguish some of the conglomeratic beds from the brecciated ignimbrites. For example, there is a passage from a pale green breccia composed of underlying volcanic rock fragments into quartz sandstone (IGS Sheet 2, 1982, map notes).

Figure 54. Dessication cracks (x 0.44).

Down slope and dip at beach level there are more obvious outcrops of the Rozel Conglomerate Formation (**Figure 53**). Here, there are dark red-brown and purple, interbedded lithic sandstones, grits and mudstones. These are discontinuous laterally and

Figure 55. Trace fossil or mud crack? (x 0.44).

occur as units consisting of grit layers passing up into siltstones and mudstones. In addition to these graded bedding structures, there are also networks of small desiccation cracks (**Figure 54**), some lines and rows of possible micro ripples and some small pits, possibly rain drop marks, on the surfaces of the red-brown mudstones. On one surface there is a narrow (5 mm), raised, right angular structure (**Figure 55**), each part being about 5 cm long. Its texture is slightly coarser than that of the mudstone and it may represent the trail of a trace fossil or an infilling of a longer mudcrack.

Above this sequence, there are discontinuous beds of conglomerate consisting of pebbles of several different Jersey rock types. Pebbles of the Jersey Shale Formation and the local rhyolite and ignimbrite formations can be found. There are also pebbles of various granites, but different from the present local granites and the foliated granite of Les Écréhous islands and reefs. The pebbles are embedded in a matrix of clay with limonite and some haematite which cause the yellow-brown and red-brown colouration.

Figure 56. Mixed conglomerate clasts (x 0.04).

Further east, up the succession, the beds still consist of pebbles of different sizes and shapes indicating poor sorting, and in the high cliffs, very large, pale coloured, rounded pebbles and boulders, 60 to 90 cm across, can be seen near the base (**Figure 56**).

The relationship, composition and structures of the two formations indicate that the ignimbrite and spherulitic flows were formed, then their upper surfaces were eroded unevenly. This irregular surface was covered by streams and ephemeral pools in which were deposited pebbles, grits and muds, possibly from flash floods. Periodic formation of desiccation features was followed by more mud deposition and then repeated, rapid deposition of unsorted pebbles and boulders. This is thought to have happened in an intermontane basin surrounded by a highland area of older Jersey rocks, the erosion of which (excluding the granites) produced a molasse deposit (IGS Sheet 2, 1982).

T5b - Belval to Le Creux au Diable

This traverse is situated on the north eastern coast of Jersey, 0.75 km south west of St Catherine's Breakwater, which represents an interval about half way up the Rozel Conglomerate Formation the base of which was seen at La Tête des Hougues.

Figure 57. Conglomerate outcrop (x 0.07).

It starts on the coastal path at its junction with the northern end of the Belval slipway and proceeds north eastwards towards St Catherine's Breakwater. The exposures form low cliffs down to the right of the path and one descends onto these after about 75 m. This part of the Rozel Conglomerate is typical of the conglomerate section; it exhibits the variety of clasts, the variation in shape and the degrees of sphericity. Several styles of bedding and different sedimentary structures can also be seen.

To the east of the footpath along the first part on the sea wall, the outcrops below are of conglomerate with incipient bedding and well defined joints (**Figure 57**). The clasts appear generally as pebbles which vary in size and are sub-rounded to sub-angular in shape with pink, igneous varieties standing out, as they are more resistant to erosion.

When one descends from the wooden steps at the end of the sea wall onto the outcrops forming the first small point, one begins to walk up the sequence across the beds from bottom to top, and down several bedding planes into gullies. Small cliffs out of the gullies show the depositional history from the bottom of the beds to the top.

On the first exposure, the bedding is massive and the clasts are clearly seen as an unsorted mixture of rock fragments from about 2 to 15 cm in the longest direction. They are sub-angular to rounded with the larger being more rounded and some show preferred orientation (**Figure 58**). Scattered pink, rounded microgranite and quartz porphyry pebbles are obvious with their longer axes parallel indicating weak stratification. Smaller, grey, variously laminated shale and volcanic rock fragments are mainly angular, their shape being controlled by close jointing in the parent rock (The Jersey Shale Formation) and apparently by little abrasion. Weak bedding indicates a dip varying from 45° to 60° to the N and NNE. The rock exhibits some conjugate joints and some faults which strike ENE, one of which may have controlled the formation of the first gully.

Figure 58. Clasts with preferred orientation (x 0.1).

Over the next section, past the gully, the still largely massive, unsorted deposit contains scattered small boulders, 20 to 30 cm in the longest direction which indicate bedding (**Figure 59**).

Figure 59. A small boulder (x 0.07)

In the next gully a zone of wedge and lens shaped beds can be seen which may represent original sedimentary structures or be beds produced by faults oblique to each other. The small cliffs on the north side of the gully have a similar arrangement of clasts but scattered larger boulders appear from half way up the cliff and continue towards the next gully. The boulders are 30 to 45 cm long and indicate a change in the depositional environment.

Figure 60. Conglomerate beds (x 0.03).

The next gully is more interesting as it reveals well defined bedding from the base of the small, northern cliff face. The several beds vary in thickness from 30 cm to 2 metres. The clearly defined bedding is a result of differential erosion along and into thin, red-brown shale beds (**Figure 60**). Some beds thicken and thin and wedge out; some show grading from 1 cm to 6 cm sized fragments and the shape of the clasts is sub-rounded in the lower two beds. In the third bed, there appears to be reverse grading from a thin, discontinuous red shale less than 1 cm thick to a coarser sandy bed, 3 to 4 cm thick, overlain by a bed of smaller sub-rounded pebbles grading into larger pebbles with scattered larger boulders 20 to 30 cm across. One such boulder is of a pale yellow brown foliated granite, similar to that on Les Ecréhous; another seems to be a grey arkose.

Figure 61. Colour change in layers (x 0.12).

The colour also changes, from dark red-brown to green-grey, where the clast size changes within the beds (**Figure 61**). The colour change boundary is irregular and represents a change from ferrous iron to ferric iron (haematite - red) due to oxidative weathering by hydrolysis. Other different rock types forming the pebble clasts include grey laminated shale, milky quartz and pink feldspar porphyry. A quartz vein dips at 60° northwards and seems to be parallel to the bedding.

From the top of this gully and cliff, north eastwards to the next cove (Le Creux au Diable), the beds are massive with unsorted small and large clasts of the same compositional variation, and still from small angular to larger rounded types (**Figure 62**). Some show preferred orientation and some rare imbrication. One or two clasts stand vertically across the bedding. Some rectangular holes occur where

Figure 62. Clast variety (x 0.08)

rectangular jointed clasts have been weathered and eroded out.

At the top of these low coastal cliff outcrops on the approach to the coastal footpath around Le Creux au Diable a slightly irregular erosion surface is overlain by glacial head of yellow sand and clay with angular clasts and several small boulders of conglomerate (**Figure 63**). They may represent part of a solifluction deposit or, possibly, a raised beach as the pebbles are rounded.

Figure 63. Glacial head with boulders of conglomerate (x 0.11).

The Rozel Conglomerate has been described as an intermontane molasse deposit in the previous site notes. This traverse reinforces that conclusion, and the various sedimentary structures, the different clast shapes and sizes indicate the varying strengths of the periodic flash floods which transported and deposited the many different rock types in this basin.

Explanatory Notes
No.6a and 6b.

Le Petit Étacquerel
to
Le Pulec,
and
L'Étacq NW Quarry

Ordnance Survey Grid References
54735465 to 54735494, and 55855433

**CARE SHOULD BE TAKEN TO CHECK TIDE
TIMES AT THE FIRST SITE.**

The duration of the first visit can be
some two hours.

A period of half an hour may be allowed for the
second site not taking into account the time taken
to get from one site to the other.

The geological sketch map T6 may be used to locate
the various points of interest by using the positions
of the Figure numbers,
e.g. Site 64 on the map refers to Figure 64.

Key to Geological Sketch Map T6

NW granite St. Mary's type

Jersey Shale Formation

Faults
Inferred faults
Inferred geological boundaries
Mica Lamprophyre dykes
Coast line
Cliff line

Rock stack

Roads
Footpaths
CP   Car Park
PH   Public House

Geological Sketch Map T6, Le Pulec and L'Étacq.

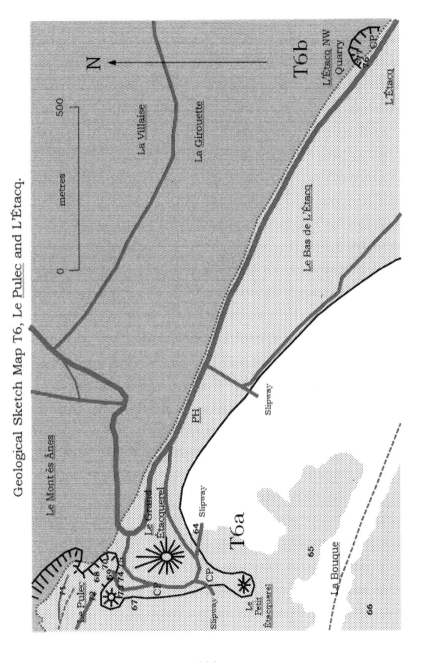

Geological Sites of Special Interest

Please note that this trail includes outcrops(s) that are protected as Sites of Special Interest (SSIs) under Jersey's Building and Planning Law (2002). Please see page 134 in this book for more details.

T6a - Le Petit Étacqerel to Le Pulec

The area is situated in the north western part of Jersey along the coast about 1 km north west of L'Étacq (558545) and is noted for its variety of rock types, structures and geological history. The rocks are well exposed in beach outcrops, cliff sections and inlets. They comprise outcrops of the Jersey Shale Formation of late Precambrian to Lower Cambrian age, 580 to 540 Ma, (Guerrot and Puecat in D'Lemos *et al.*, 1990, p.214) and the NW igneous complex, dated at 480 Ma (Bishop and Bisson, 1989, p.52) and 438 Ma (Brown *et al.* in D'Lemos *et al.*, 1990, p.196).

Figure 64. Sedimentary structures (x 0.04).

The first exposures occur on the south side at the beginning of the causeway to Le Petit Étacqerel and consist of medium grey, well bedded siltstones and interbeds of purple, fine grained sandstones (greywackes) with a variety of sedimentary structures (**Figure 64**).

Figure 65. Fold closure dipping gently south (x 0.05).

The thin interbeds are contorted and exhibit prolapse folds or slump beds, attenuated beds which thicken into lenses and possibly some sedimentary boudins. These structures indicate deposition in an unstable environment and represent turbidites which were disturbed penecontemporaneously, whilst partly consolidated on a submarine slope. Other beds, exposed around Le Grand Étacquerel exhibit sole markings of flute and bounce casts which reflect pitting on the upper surface of the underlying beds.

Other structural features can be found on the foreshore though the weed covered rocks are a hazard to progress. On the far side of the first outcrop to the south just before the large gully known as La Bouque the crest of a fold system can be seen (**Figure 65**). Across the gully vertical beds of the Jersey Shale can be found (**Figure 66**).

Figure 66 (left). Vertical bedding (x 0.01).

Figure 67 (right). Ripple bed (x 0.07).

Proceeding about 150 metres NNW either by road or foreshore, outcrops of similarly interbedded grey siltstones and pink to light purple, sandstone interbeds occur below and to the west of the small car park. Here the interbeds are 1 to 2 cm thick, between 2 to 3 cm thick beds and exhibit different sedimentary structures such as crossbedding (ripple lamination) and grading.

Some surfaces are rippled (**Figure 67**), these ripples being related to the crossbedding of current action rather than to the oscillation of wave action.

Figure 68. Le Pulec (x 0.003).

Climbing up to the car park and walking eastwards permits an elevated panoramic view (**Figure 68**) of Le Pulec Cove and its geological setting prior to descending the slipway to the cove itself. Opposite this vantage point eastwards can be seen the high cliffs of light brown-yellow to pink NW granite and its irregular contact with the medium grey, well bedded laminated siltstones of the Jersey Shale Formation.

In addition, several light coloured dykes and veins can be seen cutting across the Shale Formation and also several inlets and caves which have been eroded along faults and master joints. Thick deposits of yellow-brown glacial head form cliffs at the head of the cove and above the bed rock.

Figure 69. Beds on the slipway (x 0.008).

Descending the slipway and into the cove one walks up-sequence past a succession of grey laminated siltstone beds, some 60 cm thick, dipping at about 60° to the north-east. They are compact, granular and spotted in part, and are hornfelses in this section as a result of contact metamorphism (**Figure 69**). N.B. Faults and erosion/deposition features in this part will be described later in chronological order.

Across the cove, the eastern cliff face is an exposure of younger sedimentary beds above those of the slipway and the western beach (**Figure 70**). They consist of medium grey, bedded siltstones and light grey laminated interbeds of fine grained sandstones (greywackes) dipping generally NNE. Small scale sedimentary structures can still be seen, as can small scale faults, for example step faults and rift structures, the displacement being measured in millimetres.

Figure 70. Younger sedimentary bedsw with veins and dykes (x 0.02).

Striking across these beds are thin light coloured veins of milky quartz about 0.5 mm thick and several thicker, occasionally bifurcating, quartzo-feldspathic dykes about 15 cm thick, showing evidence of multiple intrusion by changes in colour, crystal size and concentration. Adjacent low beach outcrops show quartz veining and scattered small pyrite crystals less than 1 mm across.

Further north-west, the irregular contact with the light brown to yellow and pink granite is well defined. The granite is coarsely crystalline and consists of grey glassy quartz, white and yellow to pink euhedral feldspars with minor biotite and hornblende. There are no xenoliths of shale in the granite and the respective margins show no marked changes (except where the granite is greyer in parts), indicating that it was relatively cool when intruded.

Turning slightly northwards, up and over a low granite outcrop, the rock platform consists mainly of granite and metamorphosed siltstone (hornfels), strongly jointed and faulted. Erosion along a narrow fault zone has formed a small inlet south eastwards into the cliff. Where the fault extends across the platform it forms a long narrow gully with the floor exhibiting marked differences in colour from light brown-yellow to dark purple and bronze. This is a mineral vein, varying form 4 to 15 cm wide and consisting of zinc blende (sphalerite) and a gangue mineral of ferroan dolomite (Bishop and Bisson, 1989, p.110) and ankerite (Mourant, 1961, p.6) **(Figure 71)**.

To the north-west along the base of the cliffs a marked colour change reveals the shale formation and granite contact which can be easily examined in several places. Here too, several caves have been eroded along major joints into the granite.

Figure 71. Sphalerite vein (x 0.25).

West of this outcrop, near a track in the beach and under the gravel and sand there are two argentiferous lead and zinc veins (**Figure 72**). In the late 1800s attempts were made to mine them at low tide with some ore going to Wear Quay on the River Tamar in Devon. The venture soon failed for economic reasons.

Figure 72. Site of nineteenth century mining activity (x 0.01).

Returning to the slipway, the final exposures can be examined. Firstly, at the top of the north side there is a good example of a fault zone, about 60 cm wide, identified by the crushed nature of the rock and its yellow-brown colour due to weathering and decomposition (**Figure 73**).

119

Figure 73 (left). fault zone (x 0.02).

Figure 74 (right). Large filled pothole or gully (x 0.03).

Secondly, facing this on the south side, the shale formation outcrop has an irregular erosion surface, above which lies a thick deposit of glacial head. The north-easterly dipping 'shales', often a spotted hornfels at this location are overlain unconformably by a deposit of yellow-brown glacial head consisting of angular 'shale' fragments in a matrix of sand and grit, and exhibiting incipient

stratification. Within the rock outcrop section there are two vertical exposures of angular fragments and pebbles extending down from the unconformity to the slipway and bounded on either side by the shale. The first, opposite the
fault zone, is wide near the erosion surface and narrows downwards to the slipway (**Figure 74**). It consists of angular fragments of head overlying oval granite pebbles at the base (junction with the slip) often covered by landslip. The second is situated about 20 m down the slip and is narrower, about 1 m across (**Figure 75**). The pebbles at the base are again granite and oval and generally lie horizontally. They are overlain by the angular fragments in

Figure 75. Small filled pothole or gully (x 0.03).

a sandy matrix of head. The exposure may be a section through an infilled cylindrical pothole, or across a gully. If the latter, it may represent a pebble filled gully on a former beach platform which has been raised to its present level [similar pebble-filled gullies occur to the west in the present day wave cut platform]. They both were later filled by head, a solifluction deposit formed in glacial times.

Structurally, this traverse crosses beds which strike NW and dip NE at 45 to 76°. They are well jointed and minor faults strike NE while major ones strike between E and ESE. Further west on the wave cut platform the beds dip at >80° at some localities and some are overturned. Plunging anticlines can also be seen showing a southerly plunge indicating a second period of folding. Initial folding was due to NE to SW compression whilst a second period producing the plunging folds was from N - S tilting them to the south. Around the traverse site the folds are described as N - S striking folds (D1) modified later to E - W striking (D2) while the variation in strike is thought to outline a major easterly plunging St Ouen Anticline (Bishop and Bisson, 1989, p.73). More mapping seems needed here.

T6b - L'Étacq, NW Quarry

Finally, south at L' Étacq, a disused quarry provides confirmation of the shale-granite boundary. Here the contact between the north easterly dipping grey Jersey Shale Formation and the pink to light grey NW granite can be seen steeply inclined to the west. The contact is well defined and irregular (**Figure 76**). There is no brecciation nor any evidence of xenoliths or very chilled margins which indicates a relatively

Figure 76. Granite-shalle contact (x 0.07).

cool temperature during intrusion. The granite is well jointed but not at 90° angles. An interesting feature is the way that the joints are curved parallel to the contact and the upper surface suggesting relative tension during cooling or release of confining pressure when the overburden or country rock was eroded (**Figure 77**). Within the granite there are isolated, discontinuous veins of quartz with disseminated, flat, silvery molybdenite crystals several centimetres from the contact.

Figure 77. Inclined and horizontal jointing.

In summary, the oldest rocks are those of the Jersey Shale Formation which from their composition and structures indicate that they formed in ephemeral channels and on the fringes of a submarine deltaic fan subject to turbidity currents, at the mouth of a submarine canyon at the foot of a continental slope (Bishop and Bisson, 1989, pp.10-11). It is thought that this fan formed part of the North Cotentin arc basin during Upper Brioverian times in the north eastern Armorican massif, lying south of a southerly descending subduction zone (D'Lemos *et al.*, 1990, pp.124-125). The palaeocurrent directions cited in the two studies differ and need to be resolved. Volcanic activity followed, examples of which are included in the other Explanatory Notes, and then the formation was folded and tilted at least twice by E - W then N - S compression during the later part of the Cadomian Orogeny, possibly about 540 Ma (D'Lemos *et al.*, 1990, p.7).

The shale formation at the traverse area was then intruded by the coarse, outer part of the NW granite which formed a post tectonic complex in late Cadomian times, possibly as a volcanic arc granite, from assimilation and fractional crystallisation. Details of the geochemistry and the above interpretation are given by Brown, *et al.* (D'Lemos *et al.*, 1990, p.197). At this time the Jersey Shale Formation was thermally metamorphosed to become a low grade spotted hornfels in addition to regional metamorphism to low grade green schist facies.

The late Cadomian periods of folding were followed by periods of uplift, weathering and erosion during the Palaeozoic to produce the Rozel Conglomerate (see Explanatory Notes No.5a & 5b) during the Mesozoic and Cenozoic, sea floor deposits of those eras suggest that Jersey was surrounded by Cretaceous and Eocene seas. Although the Pleistocene ice never reached Jersey, the recent glaciations caused sea level changes, marine deposits, raised beaches, and the glacial head on top of our cliffs. Current erosion and deposition are altering the coastal features of this area.

Glossary of Selected Geological Terms

The rocks and minerals in the text are described from field examination of hand specimens using a lens; no microscopic or chemical analyses were done. The rocks are identified using the following elementary names and characteristics:

Igneous rocks Consist of crystals of different minerals which vary from light to dark in colour and from fine to medium and coarse in size, produced by crystallisation from magma or lava.

Sedimentary rocks Consist of grains (clasts) of different rock types of variable size and shape, of crystals or of fossils deposited in air or water.

Metamorphic rocks Consist of crystals of different minerals, which occur with certain structures such as cleavage or banding, produced by increased heat, compression or both.

The name given to any rock within the above groups depends on its texture and mineral assemblage. This has produced a difficult nomenclature but the relevant features will be described in the text.

The minerals are identified by their colour, crystal form (such as columnar or lath like) or their symmetry (such as cubic or hexagonal). The colours of the minerals determine the colour of the rock and also indicate the possible environment, mode of formation or the type of subsequent alteration. Their names are and derive from their appearance, from the way they break, their original locality or the name of their discoverer.

Rocks

(with reference to Allaby, A. and M., Eds. *Dictionary of Earth Sciences*, 1999.)

Agglomerate   A sedimentary rock of volcanic origin composed of angular fragments (pyroclasts) of different size and shape.

Andesite   An intermediate, volcanic igneous rock, generally medium grey, finely crystalline, but with scattered large white crystals locally.

Aplite   An acid, minor intrusive, igneous rock, red-pink in colour, of finely crystalline, uniform texture.

Appinite   A dark coloured plutonic igneous rock with light coloured plagioclase feldspar and rich in black elongated prismatic crystals of hornblende. A variety of diorite.

Conglomerate   A clastic sedimentary rock composed of rounded fragments of other different rocks, of variable size and shape in a matrix of finer grained material.

Diorite   A medium coloured, coarsely crystalline plutonic igneous rock. It contains light coloured plagioclase feldspar and black hornblende plus or minus biotite mica and augite.

Dolerite   A basic, minor intrusive igneous rock, generally dark grey and with a medium crystalline texture, which occurs in dykes.

Gabbro   A basic, major intrusive, plutonic igneous rock, dark grey=black with a coarse crystalline texture.

Glacial Head Consolidated and unsorted, composed of angular fragments of different size and shape in a matrix of sand and clay; usually forming cliff tops overlying the bedrock.

Granite An acid, major intrusive igneous rock, red-pink and light grey with a coarse crystalline texture.

Greywacke A clastic sedimentary rock, medium to dark grey, composed of poorly sorted, sand size grains of rock (lithic) fragments, quartz and feldspar in a finer grained groundmass.

Hornfels A metamorphic rock, dark grey, fine to medium crystalline texture, very compact produced by thermal alteration of a shale or mudstone.

Ignimbrite An acid volcanic rock composed of ash, pumice and glass shards often fused to give a streaky appearance; formed subaerially from a fast moving pyroclastic flow.

Mica Lamprophyre A basic, minor intrusive igneous rock, generally dark brown, with large biotite mica crystals in a finer crystalline groundmass, which occurs in dykes.

Mudstone A clastic sedimentary rock made of clay minerals, similar to shale but massive and compact not fissile (flaky).

Rhyolite An acid, volcanic igneous rock, generally red to brown locally, of a finely crystalline texture often with flow banding and spherical concretions (spherulites).

Sandstone A clastic sedimentary rock, grey to brown, usually composed of fine to coarse sand size grains of quartz, although feldspars may be present, in a groundmass of mud or crystalline cement.

Shale A clastic sedimentary rock, medium to dark grey, composed of clay minerals and fissile (flaky) in appearance.

Minerals

(with reference to H. H. Read, *Rutley's Elements of Mineralogy*, 1976.)

Actinolite Consists of green fibrous crystals of hydrated, calcium, iron, magnesium silicate produced by low-grade metamorphism of iron and magnesium silicates such as hornblende in diorites and gabbros.

Augite Consists of dark green to black, massive to granular crystals of a complex iron, magnesium, calcium, aluminium silicate in the diorites and gabbros. It can be distinguished from hornblende in transverse sections by showing 90° cleavage.

Barytes Consists of white to pink massive, dense crystals of barium sulphate in veins often as a gangue mineral to lead and zinc ores.

Biotite Mica Consists of brown to black, platy crystals of hydrated iron, magnesium potassium, aluminium silicate of magmatic origin in granites.

Calcite Consists of small, white crystals of calcium carbonate in veins within granites and gabbros.

Chlorite   Consists of green platy or granular aggregates of hydrated iron, magnesium, aluminium silicate, a secondary mineral due to alteration of biotite and hornblende minerals.

Feldspar   There are two main types: Orthoclase which generally forms large, pink to white, lath shaped crystals and is a potassium, aluminium silicate in the granites, and Plagioclase which generally forms smaller, white lath shaped crystals and varies from a calcium to sodium rich, aluminium silicate in the gabbros, diorites and granites. Some of the orthoclase crystal masses are large enough to be called pegmatites.

Hornblende   Consists of small to large, black columnar and acicular (needle-like) crystals that in hexagonal cross sections show $120°$ cleavage. It is a complex iron, magnesium, calcium, aluminium silicate found in gabbros diorites and granites.

Muscovite Mica   Consists of silver to light yellow, platy crystals and forms as a hydrated potassium, aluminium silicate in the granites.

Pyrites   Consists of small, silver-yellow cubes and granular crystals of iron disulphide found between the larger crystals in some of the gabbros, diorites, gabbros and shales.

Quartz   Consists of small to large, light grey, glassy to milk white crystals, generally anhedral (without crystal faces), between the feldspars in the granites. Some crystal masses are large enough to be called pegmatites.

Features and Processes

Assimilation   The incorporation of surrounding country rock into the magma during intrusion, which may be left as remnant masses called xenoliths.

Brecciation   The breaking up of rocks into angular fragments to form a breccia.

Dyke A minor intrusive, wall-like structure of rock, vertical or inclined, but cutting across the surrounding rock. Groups of these may form a dyke swarm.

Fault A fracture in the rocks either side of which the rock masses have moved relative to each other: one mass downwards produces a normal fault; one mass up and over the other produces a reverse or thrust fault; and one mass horizontally past another produces a tear, wrench or transcurrent fault.

Foliation A planar structure which gives a fractured, thin sheet like appearance to the rock due to separation either side of platy minerals.

Formation A uniform sequence of a rock type that is distinguishable from another rock type by specific characteristics.

Fractional Crystallisation   The separation of different crystalline mineral fractions from the magma during cooling, according to the crystallisation temperatures of the minerals.

Lapilli   Volcanic bombs in the form of small stones of pumice or other volcanic rocks.

Mantle   The zone between the earth's crust and its core.

132

Molasse   A sequence of sedimentary rocks deposited in inter-montane basins.

Orogeny   A mountain building event which deforms rocks and strata producing joints, faults and folds.

Pegmatite   A very coarse crystalline mass of different minerals, several centimetres in length, commonly of quartz and orthoclase feldspar crystals in granites and diorites.

Plutonic   A name given to rocks and structures, which form deep in the earth's crust and mantle.

Porphyry   A rock and textural term to describe large crystals in a finer crystalline groundmass.

Turbidity Current   A submarine, sediment-laden current, which flows down the continental slope and produces a variety of sedimentary structures, such as sole marks, cut and fill structures, ripple bedding, and graded bedding when the current stops and the sediment settles to form a bed or stratum.

Unconformity   A discontinuity between an underlying set of beds which dip at a different angle to an overlying set. It is produced when the lower beds are uplifted, folded or tilted and eroded during an orogeny and then submerged during later marine transgression and deposition.   It therefore represents a former erosion surface.

Xenolith   A mass of different rock within an intrusive rock; generally part of the country rock probably assimilated by an intruding magma. Graded or sharp boundaries indicate the difference in temperature between the two different rocks at the time of intrusion.

List of Geological SSIs

Since the first edition of this book 22 locations have been designated as geological Sites of Special Interest. SSIs have full legal protection and may not be hammered or otherwise interfered with.

A list of Jersey's geological SSIs is given below, those that are highlighted form part of the traverses desribed in this book and particular care should be taken around these. For more information see Nichols and Blampied (2016) or the States of Jersey website.

> South Hill
> **Belcroute Bay** - *Jersey Geology Trail*: T2
> Portelet Bay
> La Cotte de St Brélade
> St Ouen's Bay Peat Beds (scheduled)
> Le Mont Huelin Quarry
> **Le Petit Étacquerel** - *Jersey Geology Trail*: T6a
> **Le Grand Étacquerel** - *Jersey Geology Trail*: T6a
> Le Pulec
> Le Pinacle
> La Cotte à la Chèvre
> L'Île Agois, Crabbé
> Sorel Point
> Giffard Bay
> La Belle Hougue Caves
> Les Rouaux
> Les Hurets, Bouley Bay
> L'Islet, Bouley Bay
> **La Tête des Hougues** - *Jersey Geology Trail*: T5a
> La Solitude Farm
> **Anne Port Bay** - *Jersey Geology Trail*: T1b
> **La Motte, Le Nez & Le Croc** - *Jersey Geology Trail*: T3

Lightning Source UK Ltd.
Milton Keynes UK
UKOW07f1553260816

281578UK00013B/75/P